SCIENTIFIC AMERICAN | EDUCATIONAL PUBLISHING

SCIENTIFIC AMERICAN INVESTIGATES THINGS THAT LIGHT UP THE SKY

RAINBOWS

BY JENNIFER LOMBARDO

Published in 2025 by The Rosen Publishing Group
in association with Scientific American Educational Publishing
2544 Clinton Street, Buffalo NY 14224

Library of Congress Cataloging-in-Publication Data
Names: Lombardo, Jennifer, author.
Title: Rainbows / Jennifer Lombardo.
Description: Buffalo, NY : Gareth Stevens, [2025] | Series: Scientific American investigates things that light up the sky | Includes bibliographical references and index.
Identifiers: LCCN 2024032891 | ISBN 9781725352230 (library binding) | ISBN 9781725352223 (paperback) | ISBN 9781725352247 (ebook)
Subjects: LCSH: Rainbows–Juvenile literature.
Classification: LCC QC976.R2 L66 2025 | DDC 551.56/7–dc23/eng20240809
LC record available at https://lccn.loc.gov/2024032891

Portions of this work were originally authored by Kristen Rajczak and published as *Rainbows*. All new material in this edition is authored by Jennifer Lombardo.

Designer: Claire Zimmermann; Andrea Davison-Bartolotta
Editor: Jennifer Lombardo

Photo credits: Cover gui00878/iStockphoto.com; p. 5 (background) muratart/Shutterstock.com; p. 5 (gold) Fer Gregory/Shutterstock.com; p. 6 ganjalex/Shutterstock.com; p. 7 mark strommel/Shutterstock.com; p. 9 Albert Stephen Julius/Shutterstock.com; p. 11 ALEX_UGALEK/Shutterstock.com; p. 12 makasana photo/Shutterstock.com; p. 13 (bottom) Francey/Shutterstock.com; p. 13 (top) Colin Ward/Shutterstock.com; p. 15 Lukas Jonaitis/Shutterstock.com; p. 17 hwiwoo/Shutterstock.com; p. 19 Liolka/Shutterstock.com; p. 21 (background) Valerii_M/Shutterstock.com; p. 21 (chart) Dream01/Shutterstock.com; p. 21 (sun) Rolau Elena/Shutterstock.com.

Printed in the United States of America

CPSIA compliance information: Batch #CWSA25. For Further Information contact Rosen Publishing at 1-800-237-9932.

CONTENTS

Words in the glossary appear in **bold** type the first time they are used in the text.

RAIN AND SUN

Rain is an important part of our planet's weather. It helps plants grow, gives people and animals water to drink, and helps clean the air. Rain can also give us rainbows. These don't help the planet, but they're very pretty to look at!

Before people knew what rainbows were, they made up stories to explain them. Some **cultures** believed a rainbow was a bridge that let souls cross into heaven. Others thought it was the bow of one of their gods.

An old Irish **legend** says that every rainbow leads to a pot of gold that was buried by creatures called leprechauns.

FUN FACT

RAINBOWS APPEAR MORE OFTEN IN HAWAI'I THAN IN ANY OTHER PLACE ON EARTH.

UNDERSTANDING PRISMS

To learn how sun and rain make a rainbow, it's important to understand prisms. A prism is a piece of glass that's shaped like a triangle and has several **facets**. When light meets one of these facets, it "bends," or changes directions. This is called refraction. Refraction happens because light moves faster through the air than through the prism.

Scattered water droplets in the sky can act like a prism. When sunlight passes through the droplets, the water refracts.

Rain, fog, and the mist from a waterfall can all create rainbows.

FUN FACT

WATER REFRACTS LIGHT BECAUSE IT IS **DENSER** THAN AIR.

SPEED AND ANGLE

Refraction causes a beam of light to separate into the colors that form a rainbow. They separate because of the differences in their **frequency** and **wavelength**.

Each color has a different frequency and moves through the prism at a different speed. A color's wavelength affects the angle at which it bends when it's refracted. Colors with longer wavelengths, like red, bend less than blue and violet, which have shorter wavelengths. This is why the colors in a rainbow always appear in the same order.

This picture shows how light refracts and separates when it passes through a prism.

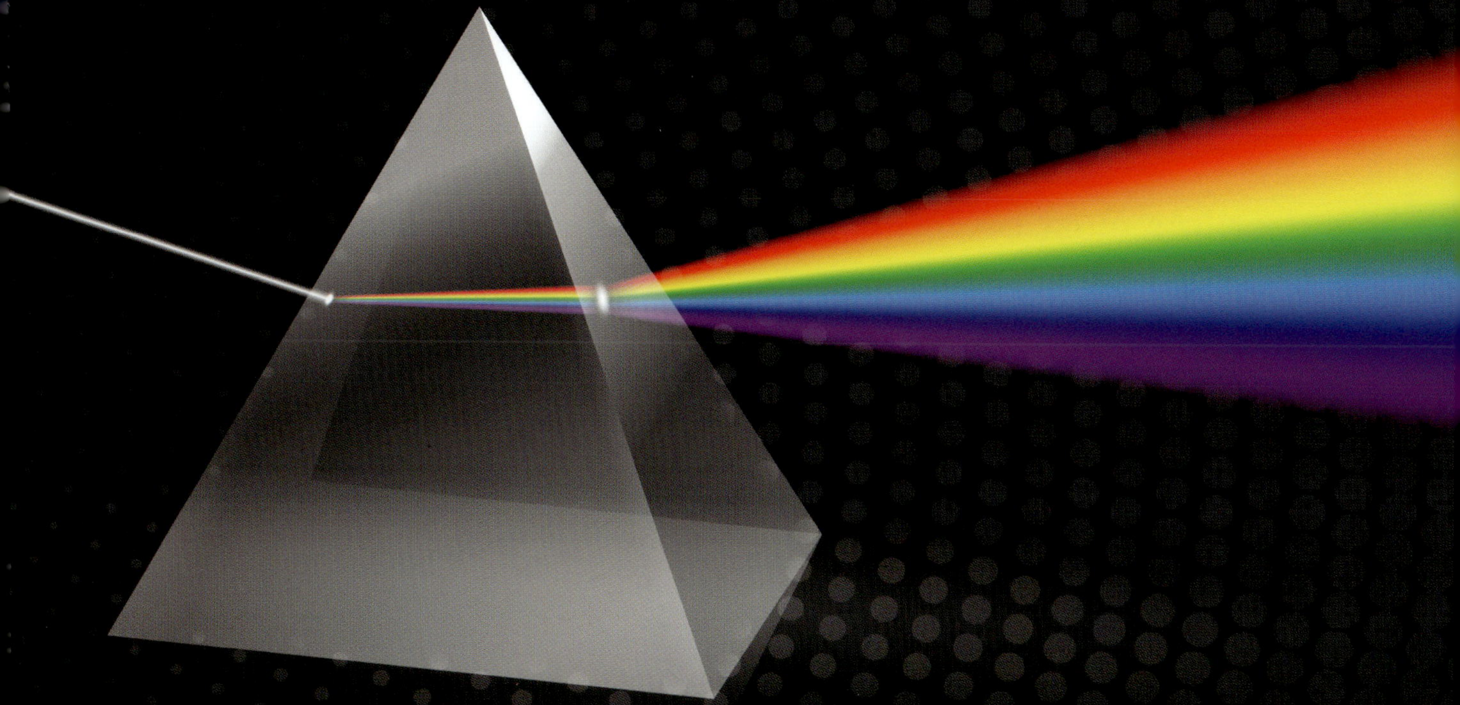

FUN FACT

THE LIGHT WE CAN SEE IS CALLED VISIBLE LIGHT. IT'S SOMETIMES CALLED WHITE LIGHT, BUT IT'S NOT WHITE! IT'S MADE UP OF COLORS.

THE RIGHT STUFF

You can only see a rainbow if all the conditions are right. The sun must be behind you, and the water droplets must be in front of you. The sun must also be at the right angle in the sky.

The formation of a rainbow begins when sunlight enters a raindrop and refracts. However, unlike a prism, the curved inside of the raindrop then reflects the light back in the direction it came from. As light exits the raindrop, it's refracted again. This creates the colored arc we see in the sky.

Because water droplets both refract and reflect light, a rainbow is actually a circle of colored light. However, the full circle can be seen only from the air. On the ground, the **horizon** gets in the way.

FUN FACT

MANY RAINDROPS ARE NEEDED TO MAKE A RAINBOW. EACH DROP SUPPLIES ONE COLOR TO THE RAINBOW. THIS COLOR IS THE ONLY ONE THAT EXITED THE DROP AT THE CORRECT ANGLE FOR US TO SEE IT.

THE RIGHT TIME

Generally, it must be raining and the sun must be shining in order for a rainbow to form. That means it's more likely for rainbows to be seen in summer than in winter, when rain turns to snow and ice.

The best time to see a rainbow is when the sun is close to the horizon. More of the full rainbow circle can be seen then. At sunset, you might be able to see a half circle of rainbow in the sky instead of just an arc.

Rainbows tend to look more like a half-circle at sunset. In the middle of the day, they often look more like a curved line across the sky.

DOUBLE THE COLOR

Sometimes two rainbows will appear in the sky at the same time. The main, or primary, rainbow is clearer and has darker colors. The second rainbow appears above the primary rainbow and is fainter.

Secondary rainbows form when light reflects twice within a raindrop. The reflections then exit the raindrop at different angles. The size of the raindrop must be just right for this to happen. This makes seeing a double rainbow very rare, or uncommon.

Because a secondary rainbow results from a second reflection of light inside the raindrop, its colors are in the opposite order of the primary rainbow's.

A PRETTY ILLUSION

Rainbows are an illusion, or a trick that light plays on our eyes. When we see a rainbow, we're not seeing anything real. This might sound confusing because the rainbow is clearly there in the sky. However, those colors are always part of the light that's all around us. A rainbow is just light that moves in a different way than we're used to.

Rainbows generally last in the sky for no more than an hour. This is because weather conditions are always changing.

A rainbow that's formed by a prism will last for as long as the light shines through the prism.

FUN FACT

THE LONGEST A RAINBOW HAS BEEN SEEN IN THE SKY AT ONE TIME WAS NINE HOURS. THIS HAPPENED IN TAIWAN IN 2017.

MOON AND FIRE

Although they're rare, rainbows can be formed by the moon. Lunar rainbows, sometimes called "moonbows," are much fainter than those caused by the sun.

Another rare kind of rainbow is called a fire rainbow. A fire rainbow forms when the sun is at a 58-degree angle in the sky and its light passes through cirrus clouds. These high, thin clouds are full of ice crystals that refract light just like a prism. The scientific name for a fire rainbow is a "circumhorizontal arc."

A fire rainbow makes clouds colorful.

FUN FACT

THE MOON LOOKS LIKE IT LIGHTS UP BECAUSE IT REFLECTS SUNLIGHT THAT SHINES ON IT IN SPACE. LUNAR RAINBOWS ARE WEAKER BECAUSE THEY'RE MADE WITH REFLECTED SUNLIGHT INSTEAD OF DIRECT SUNLIGHT.

UNIQUE RAINBOWS

Rainbows are a **unique** sight in several ways. For example, the angle from which the rainbow is observed, or seen, determines how it looks. Because no observer can look at a rainbow from the exact same angle as another observer, everyone sees a slightly different rainbow.

Rainbows are not only unique to each person but also unique to our planet. Earth is the only planet that has liquid water in its **atmosphere**, so it's the only planet where rainbows can form.

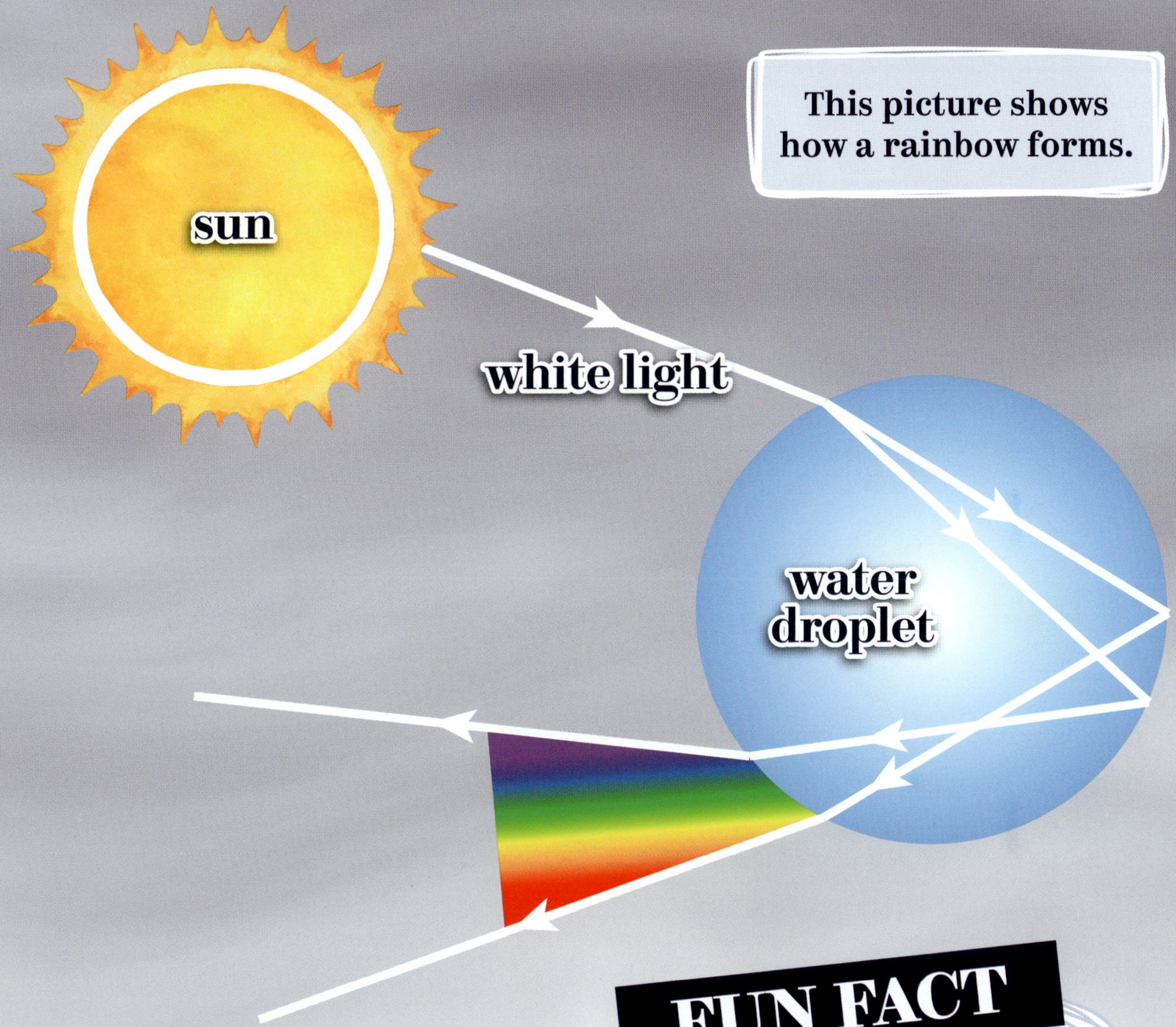

This picture shows how a rainbow forms.

FUN FACT

SOME SCIENTISTS THINK TITAN, ONE OF SATURN'S MOONS, COULD HAVE RAINBOWS AS WELL.

GLOSSARY

atmosphere: The thick mixture of gases that surrounds a planet.

culture: The beliefs and ways of life of a group of people.

denser: Packed more closely together.

facet: A small, flat area on a jewel or piece of glass.

frequency: The number of times that something (such as a wave) is repeated in a period of time.

horizon: The line that seems to form where the earth meets the sky.

legend: A story that has been passed down for many, many years that's unlikely to be true.

unique: Special or one of a kind.

wavelength: The distance between one peak of a wave and the next.

FOR MORE INFORMATION

Books

Davis, Rachael. *Over the Rainbow: The Science, Magic, and Meaning of Rainbows*. London, UK: Flying Eye, 2023.

Roehrig, Artemis. *Rainbow Science: Discover How Rainbows Are Made with 23 Fun Experiments and Colorful Activities!* North Adams, MA: Storey Publishing, 2024.

Websites

BrainPOP: Rainbows
www.brainpop.com/science/energy/rainbows
Watch a movie, play games, and take a quiz to test your knowledge of rainbows.

Kiddle: Rainbow Facts for Kids
kids.kiddle.co/Rainbow
Learn more facts about rainbows and take a look at some cool pictures.

Publisher's note to educators and parents: Our editors have carefully reviewed these websites to ensure that they are suitable for students. Many websites change frequently, however, and we cannot guarantee that a site's future contents will continue to meet our high standards of quality and educational value. Be advised that students should be closely supervised whenever they access the internet.

INDEX